INSIGHTS OF METAVERSE

DARK SIDE OF BRIGHT END

AVINASH KUMAR SINGH

Copyright © Avinash Kumar Singh
All Rights Reserved.

"This book is dedicated to the love of my life, my mother who always praised me for my efforts and better achievement."

Contents

FOREWORD

The book conveys a very clear meaning of the term metaverse. As an author I tried to give my best for this book. This short non fiction will deliever a deep intense of Insights of Metaverse. I hope you will love reading this indivisual production.

hello readers!

WELCOME TO META <3

About The Author

The author, Avinash Kumar Singh is a non fiction writer and poet of books "28 PIXELS OF FEBRUARY" and "RHYTHM OF BENJAMIN" . Avinash has an eagerness to know about life and love science. He love to explore little things which can create a big discovery in upcoming decades. The author resides in Gorakhpur, Uttar pradesh and is a one and only teenage author in the city. He says being Author was the best moment of his life cause he still can't believe that he got an Author tag in so early age.

His available books on popular platforms -

- kinda love : it hurts when it's true
- 28 pixels of february : poetries bleeds ink
- rhythm of benjamin : when alec is love

drop your valuable review after reading !

Acknowledgements

Insights of Metaverse has burried deep down and clear intentions in everyone's mind. It turned so necessary to get to know about Meta and the happenings which surrounds it. Meta has a wide atmosphere in upcoming few years. As an author I felt it is a right time to inform people why Meta is so important to learn about. This book is totally a knowledge based science book which will inform you each about Meta.

So without wasting any other seconds let's jump into the action of these data. Grab yourself a cup of tea or coffee and book a corner of peace to immerse in sense of this meta world.

happy reading moodies !

I

What is metaverse?

What is metaverse?

The metaverse is an increasingly complex idea withinside the virtual landscape, promising remarkable possibilities for billions of people. A whole definition of the metaverse is presently still being mentioned by current pioneers. However, most people talk about the "metaverse" as a brand new kind of net experience, one constructed around a number of specific technologies. According to Mark Zuckerberg, one of the marketplace leaders presently making an investment in the concept of the metaverse, the metaverse is a sort of "embodied net". It's something you may jump into (via VR) or deliver in your reality (via AR). To go a little deeper, the metaverse gives a destiny where we can experience a more potent overlap among our physical and virtual lives. Think about how the LiDAR scanner on the new iPhone can scan your environment and convey new content into them through your camera or recall how you could now put money into digital styles of art using

NFTs.

Origin of the term

The term metaverse was coined in Neal Stephenson's 1992 science fiction novel Snow Crash, in which humans, as programmable avatars, engage with each other and software program agents, in a three-dimensional digital area that makes use of the metaphor of the actual world. Stephenson used the term to explain a digital fact-based successor to the internet. Neal Stephenson's metaverse seems to its users as urban surroundings evolved along a 100-meter-extensive road, known as the Street, which spans the entire 65,536 km circumference of a featureless, black, perfectly round planet. The virtual real estate is owned through the Global Multimedia Protocol Group, a fictional part of the real Association for Computing Machinery, and is available to be offered and homes advanced thereupon. Users of the metaverse get entry to it through non-public terminals that undertake a tremendous virtual reality show onto goggles worn by the user, or from grainy black and white public terminals in booths. The users enjoy it from a first-person perspective. Stephenson describes a subculture of humans selecting to remain constantly related to the metaverse; they're given the sobriquet "gargoyles" due to their gruesome appearance.

AR - VR

AR combines real-world factors with digital augmentations. That approach digital characters and objects appear on your real-world experience via AR technology. Examples of AR which you likely didn't recognize were AR:

• Apps like Target that allow you to place furnishings in your house to see what it might look like. • Live View on Google Maps overlays the instructions over Good Street View images.

• AR engages the audience with thrilling visuals that enable creating a superior shopping experience through integrating technology with the real-life surroundings.

VR Applications

VR is now nearly intertwined into our ordinary lives. From modernizing healthcare through VR health facility tours to VR-assisted intensive army training, VR is continuously evolving. Other examples of latest and upcoming VR applications:

• VR and education guide academic games, learning, and digital hands-on experience.

• Occupational protection and health (OSH) can simulate real-life surroundings hazards earlier than an accident happens. Unlike AR, VR is a whole immersion experience for the person into the virtual experience using computer-generated simulation.

Therefore, VR calls for a headset tool that complements the alternative reality.

Growth in metaverse

The global metaverse market length was expected at USD 38.85 billion in 2021. It is expected to enlarge at a compound annual growth rate (CAGR) of 39.4% from 2022 to 2030. Major elements predicted to power the sales growth encompass a growing focus on integrating virtual and physical worlds using the Internet, growing momentum and recognition of Mixed Reality (MR), Augmented Reality (AR), and Virtual Reality (VR), and the outbreak of COVID-19, as well as the situation's next trends and outcomes. Metaverse is one of the trending era systems attracting diverse social networks and technology leaders, and online game makers to go into and establish its presence within the market.

The metaverse is a fast-growing trend with a thoughtful penetration rate of users for various programs which includes gaming, content creation, social interaction, studying and training, and online virtual shopping. According to enterprise experts, the metaverse is anticipated to infiltrate a multitude of industries in several methods in the coming years, with the potential market opportunity or the total addressable market envisioned at more than USD 1 trillion in every year revenues. The metaverse's currency is a cryptocurrency and every metaverse has its collection of coins. They're used to procuring everything from NFTs to digital real estate to avatar shoes.

Cryptocurrencies are used to attach the physical and digital worlds. They allow us to calculate the worth of digital assets in the government-issued forex in addition to the returns on the ones assets over time. The use of the metaverse to buy digital belongings using cryptocurrency is gaining traction across the world. One of the metaverse's promising possibilities is that it is massively growing to get entry to the market for clients in rising and frontier economies. The Internet has already opened up to get entry to formerly unavailable items and services. Workers from low-earnings countries, for example, may now be able to discover work in western businesses without emigrating.

Virtual fact environments will help decorate educational options, as they're a low-value and powerful manner to learn.

UI/UX

UI/UX is the art of designing interfaces and studies that make a person's virtual journey enticing and meaningful. UI and UX have grown to be the cornerstone of humanistic design in the virtual age. However, with the ever-growing variety of online structures and the rapid evolution of operational and engagement models to fulfill user needs, the future of UI/UX layout is converting dynamically. The rise of the Metaverse is beginning an entire new world, in which people see each other, interact, work and interact in a very new way.

It is a spectrum of virtual worlds and simulations where users are free to explore anything they want from building to coding and designing. With Facebook rebranding itself as Meta and greater than 160 technology leaders operating on the Metaverse, this platform will absolutely redefine advertising and interplay layout in the following few years.

The virtual reality

The technology of virtual reality has been a long-awaited dream in the tech world. The concept of a data-pushed virtual world has been around nearly as long as the idea of computing. Virtual reality itself has been a patented concept since 1961 named the Telesphere Mask. The authentic patent states that the spectator would be given a whole sensation of reality from 3-dimensional imagery with full colour and 100% peripheral vision to binaural sound, scents, and even air breezes. Today, virtual reality, VR, and headsets lack access to air breezes and scents.

The idea of virtual reality has usually been a computing dream. Even earlier than the realization of genuine VR functionality, computing fanatics drove the concept far beyond the limitations of their time. Even now, hardware add-ons are made to push those obstacles even further. With so many similarities, it's not difficult to confuse virtual reality with the metaverse. Especially because some businesses have taken to using the two terms interchangeably. VR is surely a major part of the metaverse, but it's best one part of the complete idea.

Right now, the Metaverse is a free collective of immersive studies that can be accessed with smartphones, laptops, VR HMDs (HTC Vive, Valve Index, Meta Quest), and AR Goggles inclusive of Microsoft's Hololens, desktops, and different computing devices. Most of these experiences are VR games, but there's also an extensive type of productivity suite that permits people to satisfy themselves in a virtual environment from the comfort of their personal homes.

Projects on metaverse

Gala Gala

is a blockchain gaming platform that nearly every person is conscious of. It is understood for developing a platform in which players can freely trade in-game items. This metaverse venture lays significance to the extent of authority and possession of gamers. Also, a factor to word is that the gamers can create and personalize their avatars as well.

Decentraland

Decentraland (MANA) is one metaverse blockchain venture that won recognition in no time. Right from assembling new people, to shopping, playing games, and even building new enterprise tasks, this project has it all. No surprise why this metaverse challenge will continue to grow in the days ahead.

Sandbox

Sandbox metaverse is but any other decentralized blockchain-based digital task that one should absolutely be aware of. It resembles a notably polished model of Minecraft. The sandbox permits its users to accumulate land, broaden it, trade it, and create interactive belongings utilizing the tools available.

Axie Infinity

Axie Infinity, a metaverse blockchain project, has grabbed eyeballs from everywhere with its lovable animal-like characters. Well, that's not all – the gamers additionally stand the chance to earn while gaming.

My Meta MMO

My Meta MMO is fairly famous nowadays as it has carved a niche for itself as an artificial intelligence-driven challenge in which the meta-characters farm, battle, breed, and mine. Wondering who's the mind behind this progressive metaverse blockchain project? Well, who apart from My

Meta Studio?

Bloktopia

Bloktopia is a metaverse assignment revolving around tokens. As interesting as it may get, the token holders are called Bloktopians. One component of this task can certainly not go unnoticed – this might be the primary metaverse project that enables the users to access crypto statistics and immersive content multi function place.

pax.world

pax.world is a community-owned metaverse project that has loads to do with non-fungible tokens (NFTs). A point really well worth a mention is that you may create and monetise your own content right here. Somnium Space Yet another metaverse venture that uses NFTs is that of Somnium space. Now, here is the catch – each NFT you acquire will come up with access to an exceptional 3D avatar revel in altogether.

Enjin

Wondering what makes Enjin a part of the top 10 Metaverse blockchain projects to appear out for in 2022? Well, it is the truth that it no longer contains a virtual world full of gameplay features. Additionally, it offers a platform in which creators could make NFTs easily.

Star Atlas

Star Atlas is all here to offer you an experience of the global's first AAA crypto games. Here, the 2 tokens – ATLAS and POLIS are built on the Solana blockchain. Right from a splendid storyline to monetization opportunities, this metaverse blockchain project is the entirety that desires your attention.

II
Technology and Implementation

The cutting-edge improvement of the Metaverse became viable because of technology like synthetic intelligence (AI), the Internet of Things (IoT), AR, VR, 3d modeling, and spatial and aspect computing.

Artificial intelligence

AI paired with Metaverse generation guarantees the Metaverse infrastructure's balance at the same time as additionally turning in actionable statistics for the higher layers. NVIDIA technology is an awesome instance of ways AI may be essential in growing virtual areas wherein social interactions will arise within the Metaverse.

Internet of things

While IoT will permit the Metaverse to look at and have interaction with the actual world, it'll additionally function as a three-D person interface for IoT devices, taking into account a greater personalised IoT level. Both the Metaverse and the Internet of Things will help businesses in making records-pushed judgments with minimum intellectual effort.

Augmented and VR

The concept of a Metaverse combines technology like AI, AR and VR to allow customers input the digital world. For instance, digital objects may be embedded withinside the real surroundings the usage of augmented fact era. Similarly, VR enables immersing you in a 3-d digital surroundings or 3-d reconstruction the use of three-D pc modeling. While carrying a digital fact headset or different tools isn't always required withinside the Metaverse, specialists agree with VR becoming a crucial part of the digital surroundings. However, it's crucial to notice that the Metaverse isn't like AR and VR. If you're curious to realize how you could input the Metaverse, the solution is that augmented and digital truth technology are a way to get into the dynamic 3-D virtual world.

3-d modeling

3-D modeling is a pc images method for developing a third-dimensional virtual illustration of any floor or object. The Metaverse's three-D truth is essential to making sure the consolation of its consumers.A lot of photograph amassing and image layout are required to create a 3-d world. The 3-d pics in maximum video games like The Sandbox (SAND) offer the influence that the participant is truly withinside the game. The Metaverse wishes to be constructed at the equal foundation.

Spatial and side computing

The exercise of leveraging the bodily area as a laptop interface is referred to as spatial computing. With technology just like the HoloLens, Microsoft is a pioneer withinside the subject of spatial computing withinside the metaverse area.

In contrast, aspect computing is a network-primarily based totally cloud computing and carrier shipping paradigm. Edge affords end-customers with computation, storage, facts and alertness answers like cloud computing services. To supply the identical stage of revel in as in fact, maintaining the consumer involved and immersed withinside the Metaverse is critical. In light of this, the reaction time to a person's movement must basically be decreased to a stage underneath what's detectable to humans. By web hosting a sequence and aggregate of computing assets and conversation infrastructures near the customers, facet computing offers short reaction times.

Virtual Reality (VR) has been one of the maximum captivating contributions of the technological increase withinside the beyond decade. The metaverse, the maximum good sized rising tech fashion of contemporary-day times, is ready to raise this enjoyment to the subsequent level. How approximately an immersive 3-D virtual enjoy that mixes more than one digital and bodily worlds? Well, that is precisely what the metaverse promises.

The idea is being taken into consideration the destiny generation of the net and could allow customers to meet, socialize, play games, and paintings with different customers inside three-D spaces.The term "metaverse" became conceptualized through Neal Stephenson in his technological know-how fiction novel Snow Crash, written in 1992. The novel envisaged that people may want to get away from the actual global right into a digital international called "Metaverse" with the assist of virtual avatars and discover this digital global to the fullest. Decades later, with the arrival of revolutionary technology like Augmented Reality (AR), VR, Artificial Intelligence (AI), device learning, blockchain, etc. it has come to be feasible to transform this captivating idea into reality.

Several manufacturers like Facebook, Microsoft, Nvidia, and Decentraland have began out to discover this vicinity during the last couple of years.The metaverse generation grabbed the highlight and have become a subject of eager hobby recently, while Facebook modified its logo call to Meta in October 2021 and deliberate to recognition on exploring the metaverse in full. This article speaks approximately the metaverse era in element and affords glimpses of its destiny prospects. Today, there exist numerous man or woman metaverses which have confined features. Presently, the gaming zone offers the revel in closest to the metaverse revel in, in comparison to different commercial domains. Decentraland, a start-up, created a completely unique digital global for its internet site customers withinside the year 2017. This digital global has its very own economic system in addition to currency. It integrates social factors with NFTs (representing beauty collectibles), cryptocurrencies, and digital actual estate. The gamers of this blockchain sport take part in lively governance at the platform.Microsoft released Mixed Reality clever glasses named HoloLens in 2016.

The online game Roblox additionally offers non gaming offerings like digital meetups and concerts. Facebook is withinside the manner of making a social platform powered with the aid of using Virtual Reality. Furioos, created through Unity, streams completely interactive 3-d environments in actual time. Here, the environments are rendered with the aid of using Unity's GPU server infrastructure that mechanically scales itself. Second Live gives a digital 3-D environment that is being applied for learning, socializing, and business.

This metaverse additionally offers an NFT market wherein collectibles may be swapped.

III

Major concerns

Metaverse appears to be taking many tech information channels through the storm, one may want to be aware of the thrill amplifying withinside the beyond a few months. It appears that nearly everybody is speakme approximately the Metaverse, their imaginative and prescient of the Metaverse, the Metaverse improvement or maybe the capacity of residing withinside the Metaverse. We see it acting in aggregate with such subjects like Extended Reality (XR), Machine Learning (ML), Artificial Intelligence (AI), Blockchain, Non-fungible tokens (NFT), Cryptocurrency, Hyper-practical Avatars, Digital Twin, Hybrid Ecosystems and plenty of extra. The large scope of subjects keep to purpose increasingly more confusion across the idea itself.

So earlier than we dive into the primary subject matter of this article, let's find out a chunk extra approximately the Metaverse itself. The query right here is what induced the hobby round Metaverse to surge at one of these rapid pace?First of all, the cause for this kind of speedy enlargement in hobbies is especially linked to one of the tech giants, or to be more precise, to Mark Zuckerberg's

July statement that Facebook aims to create our destiny Metaverse. The Verge stated that the imaginative and prescient revolves across the introduction of 1 unified digital universe or digital space, connecting community, products, commerce, workspace, entertainment, creators and lots greater. Even though the time period itself isn't new in any respect and has been with us considering the fact that 1992, now no longer anybody is completely acquainted with the precise meaning or definition of the Metaverse.

While specialists around the world provide their personal imagination and prescient of this unexpectedly accelerating time period, we've grown to become the Lucid Reality Labs professional and visionary to outline this ever-increasing idea. The query of private identity and illustration within reason is honest in terms of the actual world. But while talking of digital environments, or the Metaverse, one would possibly wonder, what is going to without a doubt be the factors that make up one's identity.

And most significantly the way to even show who you are, rather than some other individual or maybe not looking to mimic your existence. This is in which popularity can play an vital function in phrases of each authentication, however as evidence that the entity one interacts with is truthful and legit. The important project lies withinside the opportunity to forge facial features, pictures and voice, thus, we're sure to look at new verification techniques being evolved withinside the nearest destiny.

The idea of the Metaverse contains with it comparable social problems to that of social media platforms like Instagram, Facebook and Twitter. With impressionable humans being uncovered to a wealth of information (and this time in a greater interactive space) it may probably simplest exacerbate the issues introduced ahead through social media. Dangerous Tik Tok trends, the promotion of bad or harmful practices and lifestyles in the attempt to benefit internet clout. That's not to say the entire idea of a Metaverse is bad, but it still affords a number of hurdles society will ought to overcome. Here's a video on additional issues confronted by the Metaverse through The Infographics Show. Metaverse systems are probable to grow the cap potential variety and quantity of private facts that 0.33 events acquire. That large quantity of statistics is a goldmine tech groups and entrepreneurs can probably exploit.The extra records you placed online, the larger virtual footprint you have, so you are going to run greater risks.

Since groups may be using 0.33 events to get right of entry to and use the metaverse, leaders will must recognize what facts the ones businesses will acquire approximately their workers, clients and companions in addition to how 0.33 events will keep and use that records, Mintz said.For example, as a part of defensive the UMGC metacampus, leaders need to solution vital questions on the extent of duty in protective students' records from groups trying to marketplace and promote facts, Mintz said. For some, a number one enchantment of metaverse realities is the capacity to tackle an identification divorced from the fact of the physical, ordinary life. That excessive stage of anonymity affords possibilities for terrible actors.Research in this subject matter is in its early stages, and it is now no longer clear what mechanisms companies should use to preserve nameless avatars far from their metaverse worlds.

Anonymity already permits scams and abuse on social media and the net greater generally. Metaverse structures are in all likelihood to make that even extra widespread.

IV

Challanges in Meta

Hardware

Right now, metaverse noticeably relies upon VR (Virtual Reality), AR (Augmented Reality) and MR (Mixed Reality) technology and devices. Since maximum of those aren't lightweight, transportable or affordable, metaverse can not have a wide-scale adoption.Apart from hardware accessibility, the undertaking lies in having awesome and high-overall performance fashions that could gain the proper retina show and pixel density for a practical digital immersion.

Identity

Have you ever questioned in case your social media pal is as exciting as on-line in actual existence too? The identical element may occur with metaverse as you'll be gaining access to it thru your avatars.Another trouble lies in proving your identification as bots can without problems mimic your style, information, character and entire identification. You will want distinctive verification techniques like facial scans, retina scans, voice popularity for authentication.

Addiction and intellectual fitness

If you've watched Ready Player One, you understand precisely how metaverse can have an effect on your intellectual fitness. Addiction to the digital international won't most effectively result in intellectual fitness issues like depression, tension however reason weight problems and coronary heart troubles because of the sedentary lifestyle.

Privacy & information safety

We frequently pay attention to approximately a few MNC having a facts breach. Metaverse can store greater than your e-mail addresses and passwords. It will keep your behaviour too. With a big information mine, the generation desires to make certain records private and private statistics safe for each user. This would require new safety strategies.Currency and virtual payments Metaverse will now no longer be restricted to gaming. It could be any other online market connecting billions of customers across the international market. With such a lot of currencies and special cryptocurrencies, there could be the want for short and handy exchanges. Not to say steady transactions.

Law and jurisdiction

With social media already witnessing digital crimes, metaverse could have its percentage of lawbreakers too. Rules and rules that block an account won't be enough. You want to have right legislation.But the metaverse isn't going to exist in an actual location. It can be a digital global past global borders. That method the international locations and government want to determine their jurisdiction to make sure a secure area for the customers.

V

Trap of no exit

It won't be a twist of fate that this period, withinside the overdue 2000s, is whilst the enterprise witnessed the primary social media boom, and humanity's digital footprint have become nearly as critical as its actual ones.The difficulty over such threats to the actual international aren't totally far-fetched. Internet customers globally already spend significant quantities of time online and dedicate good sized efforts to boosting their on-line presence for each social and expert purposes.

Once the Metaverse transforms the net right into a third-dimensional digital space, may want to eliminate the want for a bodily global altogether? The metaverse does provide an inhabitable, trade surroundings to the bodily global ruled through the policies of physics of our universe, however, it's far nevertheless depending on the ones very guidelines.This is due to the fact blockchain and blockchain-primarily based totally computing nevertheless rely upon huge quantities of herbal sources and energy, that is most effective viable because of the actual global that exists round us.

Without coal or the Sun to provide electricity, minting crypto to assist blockchain ecosystems might now no longer be viable, this means that that metaverse structures might stop to exist.Currently, each different factor of the Metaverse's operations uses a few bodily aid or law. User engagement with those systems follows the policies governing our notion of mildness in order that we are able to place on VR headsets and input an immersive virtual realm. Navigation aids like far off controllers, eye-tracking, hand controllers, and omnidirectional treadmills, take advantage of human anatomy and the legal guidelines of movement to allow practical experiences.

Even in the digital global, builders mimic physics residences like gravitational anchoring to create a feel of authenticity for each experience.

On one aspect are corporations like Facebook, which are searching to govern our onramps to the metaverse—and the earnings that come from it—simply as they have got with the modern-day web. On the alternative aspect are pioneers preventing an open structure that welcomes a vast atmosphere of developers and is ruled by means of the network itself. Equipped with new era like blockchains, those newbies are preventing for a decentralized and interoperable metaverse, wherein people will have real possession of virtual property and price is shared via way of means of all who're contributing to the network.

Established corporations are deploying a big arsenal in a quest to dominate the metaverse. From Microsoft to Roblox, tech powerhouses see the metaverse as our subsequent amassing area for socializing and business—and in which human beings come together, cash is to be made. Mark Zuckerberg has declared that Facebook might be a metaverse company, now no longer a social media company, withinside the subsequent 5 years, with an aim that the metaverse attain 1000000000 humans and generate loads of billions of bucks in virtual trade via way of means of the stop of decade. The company's rebrand is even forecast to be focused at the metaverse. Over the following few years, the metaverse is predicted to take place itself usually via digital reality – an alternative, virtual international that may be used for a whole lot of private and employer purposes.

Recent high-profile bulletins through Meta Platforms (previously Facebook), Microsoft, and Sony, all endorse that headsets like Meta Quest or Sony PSVR could be the patron selections to navigate interactive and social 3-d environments.

What should we do ?

With all the buzz surrounding current metaverse traits and investments, you are probably thinking if — and how — the idea will form the destiny of social media (and social media marketing).2021 noticed a super deal of cash and assets poured into the metaverse. With systems like Meta and agencies like Nike (who currently partnered with sneaker-centric metaverse massive RTFKT Studios) making an investment big quantities of cash and assets into the metaverse, there are surely humans and groups obtainable who do suppose it's the destiny of social media.With the high-degree definitions out of the manner, let's check a few particular movements you could already carry out withinside the metaverse.

1. Network

It appears that Meta's metaverse goes to be a social platform first and foremost. After all, it wouldn't be a whole lot of a digital "reality" if customers didn't get the danger of having interaction in some manner or another.Sure, this is applicable to crypto exchanges and NFT purchases too, however it additionally includes socializing in an extra traditional sense.

2. Investment

Unless you've been dwelling below a rock for the final year, you've probably heard the terms "NFT" and "cryptocurrency." Both are critical constructing blocks withinside the metaverse and exceptional methods for customers and agencies to make investments withinside the platform.Cryptocurrency is a time period encompassing some of virtual foreign money structures. The most well-known of those are Bitcoin and Ethereum. Cryptocurrency is an unregulated virtual foreign money run through a blockchain system. Its price is in a country of quite steady flux however long time systems (specially the aforementioned ones) have skyrocketed in cost given that their inception.

One of the huge attractions with cryptocurrency is the truth that it isn't always nationalized. As such, its cost is the same in America as its miles in Japan, Brazil, and some other nations. The metaverse is an international platform. As such, cryptocurrency is the favored shape of foreign money for lots of its customers. Investing in it now seems like it's going to repay in the end as its price keeps increasing.

3. Shop

These days you could use cryptocurrency to shop for pretty much anything in actual life. Heck, New York mayor Eric Adams even widely wide-spread his first paycheck in Bitcoin and Ethereum. In that sense, the purchasing opportunities of that nook of the metaverse are endless.At the identical time, there's a shape of buying that relates some distance extra immediately to the metaverse.

Whether you're constructing your stock of NFTs or constructing your avatar's international on a platform like Roblox, there's lots of buying to do on this new digital space. We've explored the earth, the land, the seas and the oceans. We're exploring space, the moon, Mars and beyond. We're exploring our minds and our imaginations, and from our preference to go beyond limits, we're inventing new universes to explore.When it involves the metaverse, extraordinary faculties of questioning are defining it. From this perspective, the arena is cut up between:

1. Those who embody a completely digital fact in which the human escapes its physical obstacles and will become simply a moral sense related to others through fantastically superior technologies.

2. Those who assist a combined fact, a better model of what we're dwelling today. As some distance as I am concerned, I am a part of the second one organization and assume that now, extra than ever, we must pay extraordinary interest to each improvement step we take.To start, I'll outline the metaverse as a brand new net—very wonderful from Web 3.0, constructed round blockchain. This new net is momentarily supplying many unanswered questions: Will it's decentralized? How will it evolve? What will it permit us to do? When will it's

completely immersive, permitting us to attach like in no way earlier than in a virtual space?Building the metaverse ought to be finished properly, with ethics in thoughts and policies in place. I trust factors that might make sure the upward push of a good, precious and worthwhile metaverse for everyone.

VI

Impact on Creators

Writing is sort of not possible without a network. We want others to percentage the burdens and frustrations of writing; to have a good time and win; to stroll this unpredictable adventure collectively. Think of the Modernist writers in Paris like Ernest Hemingway and the Fitzgeralds and the way their network changed in generating their undying works. Their early years collectively helped form the literature we realize today.Writers have bodily groups they may be part of however as many depart the towns for extra cheap dwelling, post-COVID-19, now no longer anybody receives entry to influential and excessive-acting groups.

Virtual answers provide a fee however it's a stretch to mention they satisfy each want. The power you sense whilst you're withinside the identical room with others is tough to imitate on a 2D Zoom video.The fine instance of the way a Metaverse modifications the want for bodily network is on-line gaming. For folks who played video games in center faculty or excessive faculty, with effective imaginations, we are able to consider the friendships we made with humans everywhere in the world at the same time as gambling on-line. These reports have been pumped with adrenaline and felt so actual that truth regularly blurred past due into the night.We will enjoy a comparable enjoy as we grasp out with different virtual writers in a bodily dimension.

Writers can completely explicit themselves via their movements, facial expressions, and different micro-social signals.With virtual belongings to be had with inside the Metaverse, we are able to have alternatives to customise our area, use equipment to completely explicit ourselves, and construct our international — one which draws like-minded creators.You can meet together along with your pal in NYC, London, and Santiago, all on the identical time in a digital dwelling room decorated together along with your artwork series at the walls. Location will now no longer be an impediment for a thriving international network of virtual writers.Digital networking will remodel for the primary time in years. Traditionally, you community via means of touring a chamber of trade or a convention for your industry. The previous couple of a long time modified with the internet, social media, and industry-associated platforms. You meet a person online and ship them a right away message, likely main to a video name.The on-line-to-video answer remains a substitution frequently requiring a couple of video names to get the identical electricity, results, and impact you'll get with one espresso assembly, absolutely present.

The Metaverse gives a virtual area to bodily engage with others. Imagine assembling a writer "down the street" for your virtual global. You run into every difference due to the fact you've exceeded a not unusual place network recognized by the means of writers. You take a seat down someplace and chat, speaking via voice and frame language.Digital writers will appeal to like-minded contacts via their on-line expression. Imagine assembly in an area created through you. The new buddy notices a library withinside the room. On one shelf, you've got your preferred books — high-quality verbal exchange starters.

On the other hand, you've got all your writings offered as property. Your ebooks, articles, insightful comments, and tweets, positioned in a handy region to browse and study.The equal private library may be a gateway for purchases. You might be capable of promoting specific articles or virtual books through this spot. It will even offer unforgettable reports whilst you could go to your preferred author's area. Imagine striking out in Stephen King's library.The Metaverse will deliver writers and creators complete energy to connect to others, specific themselves, and sell their creations.Digital writers will create immersive, interactive reports withinside the Metaverse. One early instance of what this may seem like is thru Holloway, the writer innovating the web analyzing enjoy.

Holloway designed an easy-to-examine on line ebook platform. When readers purchase a e- book, they could get admission to a package of sources. For instance, if a reader buys Stop Asking Questions through Andrew Warner, they get the e- book, audio clips of sensible applications, 2,000 top rate interviews, video tutorials, and a web course. When a reader purchases the e- book, they get a deep enjoyment of the author's world and his subject.The Metaverse will provide us with greater equipment to do this. Imagine a person shopping for the e- book. They can input a room (whilst studying) in which different shoppers hold out and talk about it. Occasionally, they may get a wonder from you. You can "signal copies", routinely minting them as NFTs. As they stroll around, they get the right of entry to all of the assets covered with the purchase.If you're a fiction writer, your room may be themed with the ee-e book. You should create your mini virtual global. Imagine readers reenacting the exceptional scenes withinside the ee-e book with the characters withinside the room.

The Metaverse gives a basis to make innovation like this possible. Digital writers can price top class costs for greater fee.As the Metaverse grows, writers will see extra possibilities. Many we haven't expected yet. One concept that involves thoughts is writing scripts withinside the Metaverse.The fundamental instance we've of a virtual universe is video games like Fortnight or Minecraft. Users can construct their worlds and have interaction with human beings everywhere in the world.We can expect that the Metaverse will honor its unique suggestion via its creators.

We'll see video games and occasions withinside the global. Games that invite gamers right into a tale would require writers.Since the Metaverse could be on hand to all, sport designers don't need to depend on a handful of manufacturers to simply accept their video games for consoles. They can without problems host them withinside the universe, sparking a destiny renaissance in gaming. More video games with testimonies would require greater writers.We'll see many different possibilities because the Metaverse develops.These are some matters to pray for in a Metaverse. We have a protracted manner to move however virtual writers can put together for a brand new generation withinside the craft.

The fiction -

Technology and fiction have long shared a symbiotic relationship. Just as writers dreamed up fantastical worlds primarily based totally on imagined technologies, the ones equal worlds have stimulated engineers, technologists, and scientists—spurring breakthroughs in addition to thorny philosophical questions on their work.

VII

Feasiblity in Meta

If you ask Meta, or its peers, whether or not the metaverse is possible, the solution is confident: Yes—it's only a matter of time. The demanding situations are vast, however generations will conquer them. This can be proper of many troubles going through the metaverse: Better displays, extra touchy sensors, and faster customer hardware will show the key. But now no longer all troubles may be triumphed over with upgrades to the present generation. The metaverse might also additionally discover itself certain through technical limitations that aren't without problems scaled via means of piling bucks towards them.

The imaginative and prescient of the metaverse driven through Meta is an absolutely simulated "embodied internet" skilled via an avatar. This implies a practical enjoy wherein customers can circulate via area at will and choose up gadgets with ease. But the metaverse, because it exists today, falls some distance briefly. Movement is confined and gadgets hardly ever react as expected, if at all.Louis Rosenberg, CEO of Unanimous AI and a person with an extended record in augmented fact paintings, says the motive is simple: You're now no longer clearly there, and you're now no longer actually shifting."We human beings have bodies," Rosenberg stated in an email. "If we have only a pair of eyes on an adjustable neck, VR headsets could make paintings great. But we do have bodies, and it causes a hassle I describe as 'perceptual inconsistency.' "Meta often demonstrates an instance of this hassle—buddies surrounding a digital desk. The company's press substances depict avatars fluidly shifting round a desk, status up and sitting at a moment's notice, interacting with the desk and chairs as though it have been a actual, bodily surface.

Developers can try and repair the hassle with collision detection that prevents your hand from shifting via the desk. But remember—the desk isn't there. If your hand stops inside the metaverse, however it keeps to transport in fact, you could experience disorientedness. It's a piece like a prankster yanking a chair from under you moments earlier than you take a seat down down.Meta is running on EEG and ECG biosensors which would possibly let you flow inside the metaverse with a thought. This ought to enhance a variety of motion and prevent undesirable touch with actual-global items even as shifting in the digital area. However, even this can't provide complete immersion.

The desk nevertheless does now no longer exist, and you continue to can't experience its surface.Rosenberg believes this may restrict the ability of a VR metaverse to "brief period activities" like gambling a sport or shopping. He sees augmented fact as an extra snug long-time period solution. AR, in contrast to VR, augments the actual global in preference to developing a simulation, which sidesteps the trouble of perceptual inconsistency. With AR, you are interacting with a desk that's surely there.Figuring out a way to translate our bodily paperwork to digital avatars is one hurdle, however even though it's solved, the metaverse will face every other issue. Moving information among customers heaps of miles aside with very low latency.

VIII

Readers as our real hero

I spend a number of time in front of my telecel smartphone studying for studies. But while I'm studying for pleasure, I typically clutch a print book.I've observed that my studying is pretty exceptional whilst I'm on-line. I skim the textual content quickly, seeking out key phrases that would relate to what I'm learning instead of settling in for a protracted study. I regularly print on-line articles in order that I can examine them in tough replicas due to the fact that I discover that they are less difficult to pay attention to. For our college students who're developing in a virtual international with all its blessings and distractions, I questioned what studying practices have evolved to address the web internationally and what their results may be. The online global is sizable and there's no signal of facts advent slowing down.

Our virtual revel in is better via means of media-wealthy content material and short hyperlinks to different sites, supplying convenience, flexibility of approach, and regularly less expensive prices than print materials. We have on the spontaneous know-how of globalwide occasions and everyone's response to them and can, in turn, immediately react and make contributions ourselves.But now no longer all of this facts is impartial or maybe applicable to our needs, and the velocity at which activities are pronounced offers us little time to assess sources, assume severely or have interaction in taken into consideration reflection.In Naomi Baron's 2017 article, Reading in a virtual age, her evaluate of associated studies covered a 2011 have a look at with the aid of using Ackerman and Goldsmith.

This has a look at how one college student had a choice, they spent much less time on virtual studying, and had decreased comprehension scores. Schugar etal (2011) found that contributors analyzing on-display screens used fewer look at techniques including note-taking.

Baron's article additionally mentioned extra-current studies with the aid of Kaufman and Flanagan (2016) that discovered that scholars studying digitally did properly on answering concrete questions. However, the ones analyzing in print did higher on summary questions desiring inferential reasoning.In Baron's very own studies among 2013 and 2015 of extra than four hundred college college students from 5 countries, 86% desired studying longer texts in print and 78% whilst studying for pleasure, with 92% pronouncing it become simplest to pay attention whilst studying print. 85% of American college students have been much more likely to multitask in web surroundings and most effective 26% while studying print.

IX

Metaphor of Metaverse

The definition of the metaverse is in steady flux. Many talk over it as a shared, continual virtual area for meetings, games and socializing. Avatars, regularly cool animated film-like 3-d figures, accumulate in digital rooms, have meetings, shop, play and leave. Others see the metaverse as a layer on pinnacle of the prevailing internet, a fixed of increasing protocols allowing interconnection among apps and platforms. It's uncertain if there will be an unmarried metaverse ("the metaverse"), more than one metaverse ("a metaverse") or an aggregate of each. Maybe it is a first-class idea as a metaphor for the internet's chronic extrade.David Chalmers, a truth seeker at New York University, has been considering the character of digital lifestyles because of the unique Matrix movie.

Back in 2003, Chalmers wrote a philosophical essay about The Matrix as a part of the movie's respectable website. It's brought about his deeper paintings on simulations and digital worlds many years later.In his new book, Reality+, Chalmers argues that digital worlds may be as actual as our normal truth, a end he is reached whilst spending the pandemic hopping inside and out of them to fulfill with friends.Creating the metaverse -- Chalmers sees simply one -- calls for connecting those digital worlds in a significant manner. "Community is without a doubt critical for constructing significant social worlds, absolutely constructing groups wherein humans sense invested, like we are constructing something right here," says Chalmers, who does not see those kinds of significant connections but in current open-international social metaverse apps like Decentraland or Horizon Worlds. Behind the ones connections, of course, are massive companies. And that issues Chalmers. Since Zuck rebranded Facebook, many human beings have begun to look at the metaverse as an extension of the social network, Chalmers says, an affiliation that incorporates a record of troubling privateness and societal baggage. "But," he says, "it is now no longer that smooth to look precisely at what the opportunity is proper now."

The huge philosophical questions across the metaverse are captivating however in the end educational if human beings are not having access to it. To get to the metaverse, at the least the deluxe model that has humans excited, you want hardware. Right now the hardware is clunky and expensive.Cher Wang desires to do something positive about that. She's the CEO and chairwoman of HTC Group, the large Taiwanese hardware producer that makes the Vive VR headset, the largest challenger to Meta's Oculus as a gateway to the immersive revel in everyone's speaking approximately.The first Vive headset seemed in 2016, the identical yr because the first new release of the Oculus headset, which changed into known as the Rift. Though they endure a bodily resemblance -- they are each heavy hoods that cowl your eyes and relax uncomfortably for your nose -- the 2 portions of hardware are stepping into unique directions.

While Meta pushes goggles for the masses, HTC has been looking to generate extra traction with businesses.It's affordable to surprise if the use of present generation is a stroll lower back from the immersive imaginative and prescient of the metaverse. But I've located the most exciting thing about Meta's Horizon Workrooms app, which seems like a cool animated film convention room, is the manner it merges its global with my real bodily desk. It's an early imaginative and prescient in which blended fact may want to take the destiny of an immersive generation.

The mapping, which could me paintings with my bodily pc in digital fact, additionally makes the chair I'm sitting in sense like it is a part of the digital area. I have not finished it regularly, however after I communicate with a person in the workroom, I feel like I'm making eye contact, like I'm having an actual meetup. Even if we are rendered as cartoons. As surprising as they may be, those moments include severe drawbacks for regular paintings and comfort. It's why the metaverse now seems like it is stuck among destiny and the beyond. Just like laptops failed to die while telephones have become essential, we will likely nonetheless be the usage of telephones and laptops although AR glasses grow to be mainstream.

You see, the Metaverse is clearly a metaphor for the inexorable alternate that continually takes place whilst we create new gear that make bigger our bodily or intellectual colleges via a technological medium.So right here we're today, fish swirling in a vortex of outcomes due to algorithms, cellular computing, and decentralization. We're all turning wiser even as something is profoundly converting withinside the manner that we relate to at least one another. Looking to the beyond to make sense of the existing isn't always most effectively essential to figure the consequences of our presently shifting, invisible media environment, however a reminder that there's constantly something large and extra encompassing to explain wherein we're and wherein we're going. In different words, the human questions are triumphant over the technical, and they may be timeless.

END WORDS

THANK YOU FOR BEING A PART OF OUR METAVERSE. WE HOPE YOU LOVED READING THIS AND DISCOVERED A LOT. GIVE YOUR REVIEWS AND SUBMIT YOUR FEEDBACK ON :

    gmail - wordsarephenomenal@gmail.com
    website- thewavywords.com
    instagram - phenomenal.curves
    STAY SAFTE <3